FLYING GIANTS OF DINOSAUR TIME

BY **"DINO" DON LESSEM**

ILLUSTRATIONS BY **JOHN BINDON**

L LERNER PUBLICATIONS COMPANY / MINNEAPOLIS

To Peter Wellnhofer and Kevin Padian, prolific pedagogues of the Pterosauria

Text copyright © 2005 by Dino Don, Inc.
Illustrations copyright © 2005 by John Bindon
The photographs in this book appear courtesy of: © Sinclair Stammers/Photo Researchers, Inc., p. 13; Raymond Rye, Smithsonian Institution, National Museum of Natural History, pp. 16–17; © Luis Chiappe, Natural History Museum of L.A., p.24.

This book is available in two editions:
Library binding by Lerner Publications Company, a division of Lerner Publishing Group
Soft cover by First Avenue Editions, an imprint of Lerner Publishing Group
241 First Avenue North
Minneapolis, MN 55401

Website address: www.lernerbooks.com

Library of Congress Cataloging-in-Publication Data

Lessem, Don.
    Flying giants of dinosaur time / by Don Lessem ; illustrations by John Bindon.
        p.   cm. — (Meet the dinosaurs)
    Includes index.
    ISBN: 0-8225-1424-9 (lib. bdg. : alk. paper)
    ISBN: 0-8225-2622-0 (pbk. : alk. paper)
    1. Birds, Fossil—Juvenile literature. I. Bindon, John, ill. II. Title. III. Series: Lessem, Don. Meet the dinosaurs.
    QE871.L47 2005
    567.918—dc22   2004017918

Manufactured in the United States of America
1 2 3 4 5 6 - DP - 10 09 08 07 06 05

TABLE OF CONTENTS

MEET THE FLYING REPTILES

WELCOME, DINOSAUR FANS!

I'm "Dino" Don. I LOVE dinosaurs and all the strange creatures that lived in dinosaur times. Pterosaurs are among the strangest animals of all. These flying reptiles grew as big as airplanes! Come meet some of these amazing giants.

DSUNGARIPTERUS (DZUHNG-guh-RIHP-tehr-uhs)
Wingspan: 10 feet
Home: eastern Asia
Time: 120 million years ago

EUDIMORPHODON (YOO-dy-MOHR-fuh-dahn)
Wingspan: 2 feet
Home: southern Europe
Time: 220 million years ago

ORNITHOCHEIRUS (ohr-NITH-oh-KY-ruhs)
Wingspan: 36 feet
Home: western Europe
Time: 125 million years ago

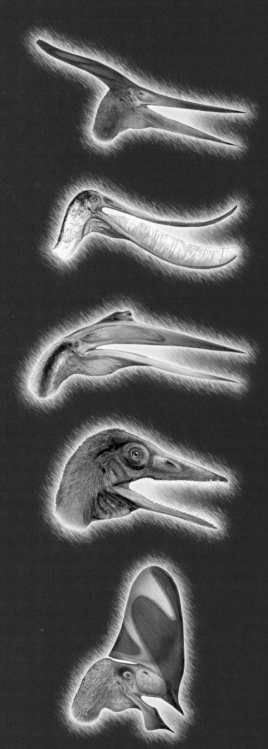

PTERANODON (teh-RAN-uh-dahn)
Wingspan: 23 feet
Home: central North America,
            northwestern Europe
Time: 70 million years ago

PTERODAUSTRO (tehr-uh-DAW-stroh)
Wingspan: 4 feet
Home: South America
Time: 140 million years ago

QUETZALCOATLUS (KEHT-suhl-koh-AHT-luhs)
Wingspan: 40 feet
Home: western North America
Time: 65 million years ago

SORDES (SOHR-deez)
Wingspan: 1.5 feet
Home: central Asia
Time: 145 million years ago

TAPEJARA (tah-pay-ZHAHR-uh)
Wingspan: 30 feet
Home: South America
Time: 120 million years ago

THE WORLD OF FLYING REPTILES

A *Pteranodon* swoops over a shallow sea in the middle of North America. In the water swims a huge sea creature. *Pteranodon* battles the sea creature for a fish.

Pteranodon was one of the biggest flying reptiles. Its wings were as wide as the wings of three eagles flying side by side. But *Pteranodon* weighed less than 40 pounds. Since it was so light, it was a fast and graceful flier.

THE TIME OF THE FLYING GIANTS

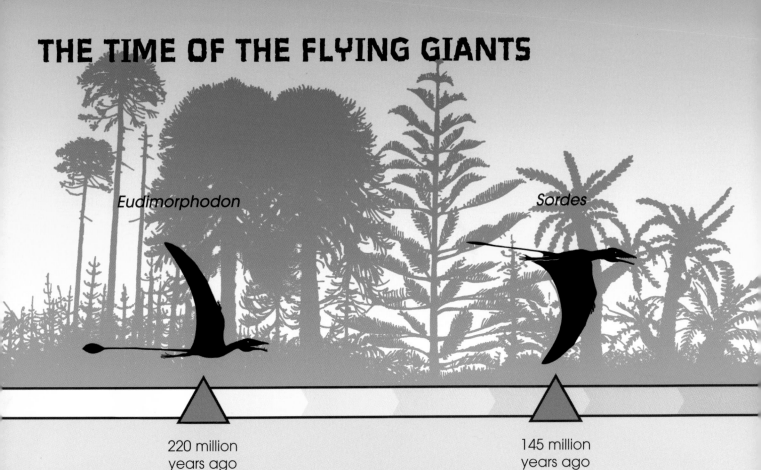

Eudimorphodon

Sordes

220 million
years ago

145 million
years ago

Pteranodon is part of a group of flying reptiles called **pterosaurs**. Pterosaurs lived millions of years before dinosaurs did. They died out when dinosaurs did, 65 million years ago.

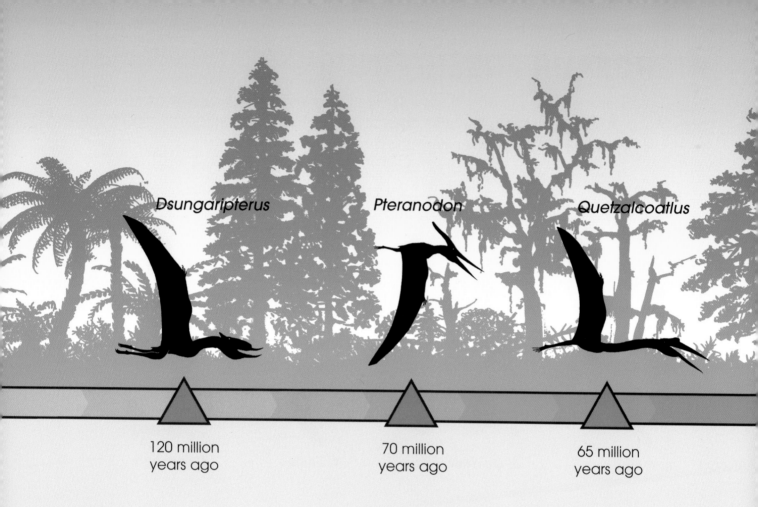

Dsungaripterus

Pteranodon

Quetzalcoatlus

120 million
years ago

70 million
years ago

65 million
years ago

Like lizards and other reptiles, pterosaurs
laid eggs and had scaly skin. Pterosaur
wings were made of stretched skin that
looked like the sails on a boat.

PTEROSAUR FOSSIL FINDS

The numbers on the map on page 11 show some of the places where people have found fossils of the pterosaurs in this book. You can match each number on the map to the name and picture of the pterosaurs on this page.

1. Dsungaripterus   2. Eudimorphodon   3. Ornithocheirus   4. Pteranodon

5. Pterodaustro   6. Quetzalcoatlus   7. Sordes   8. Tapejara

Pterosaurs flew all across the world in dinosaur time. When some pterosaurs died, they left behind remains called **fossils.** Pterosaur fossils have been found on every continent except Antarctica. Fossils of wings, heads, and tails help scientists understand how pterosaurs looked and lived.

We have found fossils of more than 100 kinds of
pterosaurs. Scientists think there may have
been thousands more. But there are not many
of these fossils. Pterosaur bones were thin and
fragile. They did not keep well as fossils.

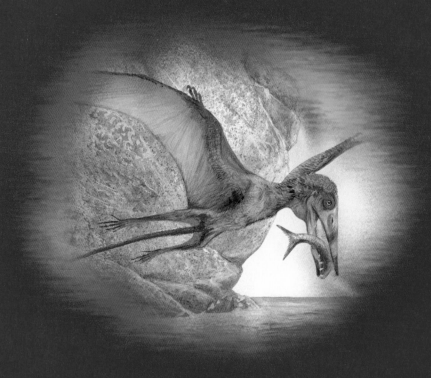

WINGS, NECKS, AND TAILS

Pterosaurs were first found in Europe nearly 100 years ago. Scientists named the pterosaur *Pterodactylus*, which means "flying finger." *Pterodactylus* had a very long fourth finger on each arm. This finger supported the wing.

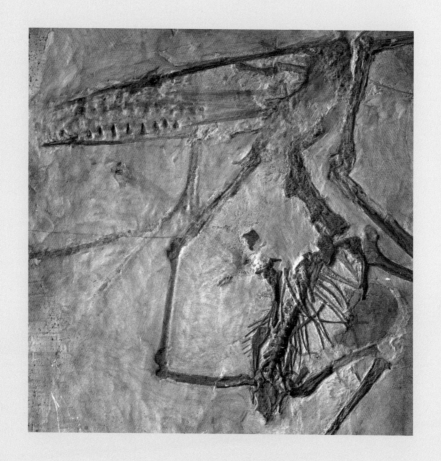

Birds have feathers to catch the wind and help them fly. *Pterodactylus* and other pterosaurs did not. But fossils show that pterosaurs had hollow bones that were as thin as straws. Hollow bones are light. They made pterosaurs light enough to fly.

These small *Eudimorphodon* are gliding and flapping over the sea. They are searching for fish that swim near the surface of the water. When they see a fish, they dive. They tilt their wings and their kitelike tails to turn.

The first flying pterosaurs lived 230 million years ago. Like *Eudimorphodon*, these little fliers had short necks and a long tail. At the end of the tail was a flap shaped like a diamond. This tail may have helped them to turn while flying.

By the middle of dinosaur time, 150 million years ago, a new and larger kind of pterosaur ruled the skies. These fliers are called **pterodactyls**. They had long necks and short tails. Their wide wings were so powerful that they didn't need long tails to help guide them.

The largest pterodactyl lived 65 million years ago. Its name was *Quetzalcoatlus.* Its enormous wings spanned 40 feet. That's longer than a school bus!

THE LIVES OF PTEROSAURS

Like birds, pterosaurs probably hatched from eggs laid in a nest. Pterosaurs might have made their nests on the sides of cliffs.

These *Ornithocheirus* parents are bringing fish to their young. The parents probably swallowed the food first. Then they spit the fish into the mouths of their hungry babies. Yum!

19

These young *Tapejara* are hungry. They are circling over the water when they see a patch of little waves. It is a sign that a school of fish is nearby.

These pterosaurs scooped up fish in their broad beaks, as a pelican does. Other pterosaurs used their sharp teeth or narrow jaws to spear fish and other sea creatures.

Some pterosaurs could catch and eat other foods. *Dsungaripterus* had a strange beak shaped like a shoe. What was the purpose of this strong beak? Scientists think that *Dsungaripterus* might have used its big beak to snatch crabs and other sea creatures that had hard shells.

This flying giant would snap its big jaws shut and crush the crabs' shells. Inside each shell was meaty food. *Dsungaripterus* would chew it for dinner or perhaps swallow it whole.

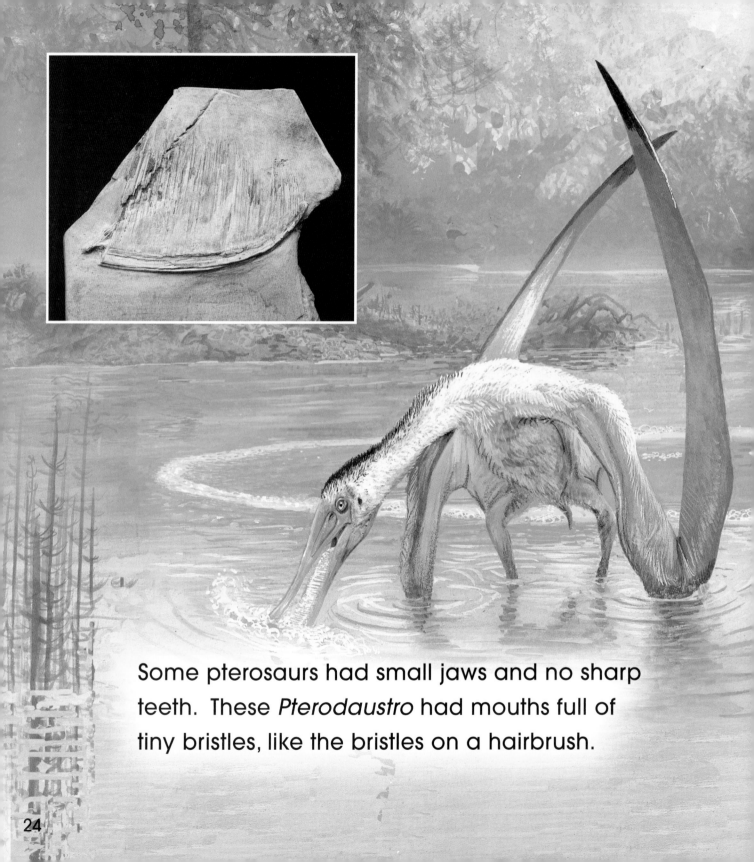

Some pterosaurs had small jaws and no sharp teeth. These *Pterodaustro* had mouths full of tiny bristles, like the bristles on a hairbrush.

Pterodaustro sucked in water and the tiny animals that lived in it. The bristles in the pterosaurs' mouths trapped the small creatures for the fliers to eat. The mouths of humpback whales work the same way.

Pterosaurs had very light bodies. Lightweight animals can get chilly easily. Many have hair or feathers to keep them warm. Scientists have discovered that some pterosaurs had hair.

This *Sordes* pterosaur is sitting in the sun and drying its coat of hair. *Sordes* was smaller than a hawk. When diving, flying through the wind, or perching in the rain, this pterosaur might have gotten very cold. A coat of hair would have helped *Sordes* stay warm.

MYSTERIES OF THE FLYING GIANTS

How did pterosaurs fly? For many years, scientists thought pterosaurs glided from trees and cliffs and caught the wind with their wide wings. More recently, some scientists thought that pterosaurs could run fast enough to take off from the ground.

Then fossils of pterosaur footprints were found.
The footprints suggest that pterosaurs could not
have run fast enough to jump into the air and
fly. We still don't know for sure how they flew.

Pterosaurs lived before dinosaurs and ruled the skies all through dinosaur time. But pterosaurs died out with the dinosaurs about 65 million years ago. What happened to these flying giants?

Perhaps changes in the weather caused pterosaurs to die. What caused the weather to change? Scientists think volcanoes erupted or an **asteroid** hit Earth. The smoke and dust would have blocked sunlight and changed the weather. Only fossils remain to show us that giant reptiles once sailed across the sky.

GLOSSARY

asteroid: (AS-tur-oyd): a large, rocky lump that moves in space

fossils (FAH-suhlz): the remains, tracks, or traces of something that lived long ago

pterodactyls (TEHR-uh-DAK-tulz): flying reptiles from the time of dinosaurs. Pterodactyls were often large fliers with short tails.

pterosaurs (TEHR-uh-SAWRZ): flying reptiles that lived before and during dinosaur time

INDEX